ACCIDENTAL
SCIENCE DISCOVERIES

MATCHES

Kenny Abdo

Fly!
An Imprint of Abdo Zoom
abdobooks.com

abdobooks.com

Published by Abdo Zoom, a division of ABDO, P.O. Box 398166, Minneapolis, Minnesota 55439. Copyright © 2024 by Abdo Consulting Group, Inc. International copyrights reserved in all countries. No part of this book may be reproduced in any form without written permission from the publisher. Fly!™ is a trademark and logo of Abdo Zoom.

Printed in the United States of America, North Mankato, Minnesota.
102023
012024

Photo Credits: Alamy, Getty Images, Shutterstock
Production Contributors: Kenny Abdo, Jennie Forsberg, Grace Hansen
Design Contributors: Candice Keimig, Neil Klinepier, Colleen McLaren

Library of Congress Control Number: 2023938019

Publisher's Cataloging-in-Publication Data

Names: Abdo, Kenny, author.
Title: Matches / by Kenny Abdo
Description: Minneapolis, Minnesota : Abdo Zoom, 2024 | Series: Accidental science discoveries | Includes online resources and index.
Identifiers: ISBN 9781098284107 (lib. bdg.) | ISBN 9781098284824 (eBook) | ISBN 9781098285180 (Read-to-Me eBook)
Subjects: LCSH: Matches--Juvenile literature. | Industrial design--Juvenile literature. | Serendipity in science--Juvenile literature. | Inventions--Juvenile literature. | Discoveries in science--Juvenile literature.
Classification: DDC 500--dc23

TABLE OF CONTENTS

Matches 4

The Accident 6

The Discovery 10

The Footprint 20

Glossary 22

Online Resources 23

Index 24

MATCHES

The discovery of fire allowed **civilization** to advance. Accidentally creating match sticks has struck up many **innovations** throughout history!

THE ACCIDENT

In 1826, English **chemist** John Walker was trying to create a new type of explosive for **mining**. A paste he had mixed with a stick **ignited** when he scraped it across the floor.

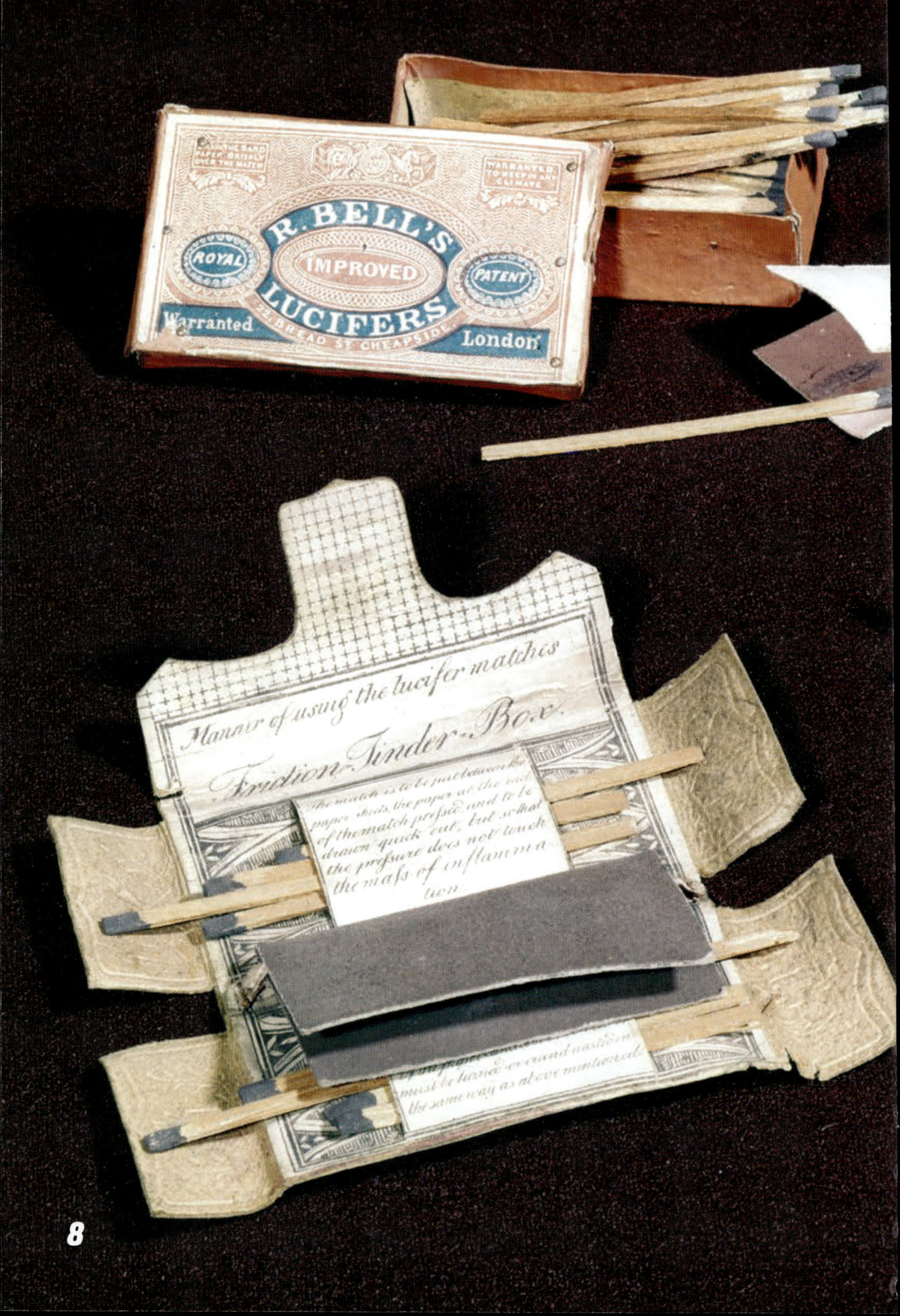
R. BELL'S
IMPROVED
LUCIFERS
ROYAL
PATENT
Warranted
London
Manner of using the lucifer matches
Friction-Tinder-Box.
of the match presse'd and to be
drawn quick out, but so that
the pressure does not touch
the mass of inflamma
tion
the same way as above mentioned

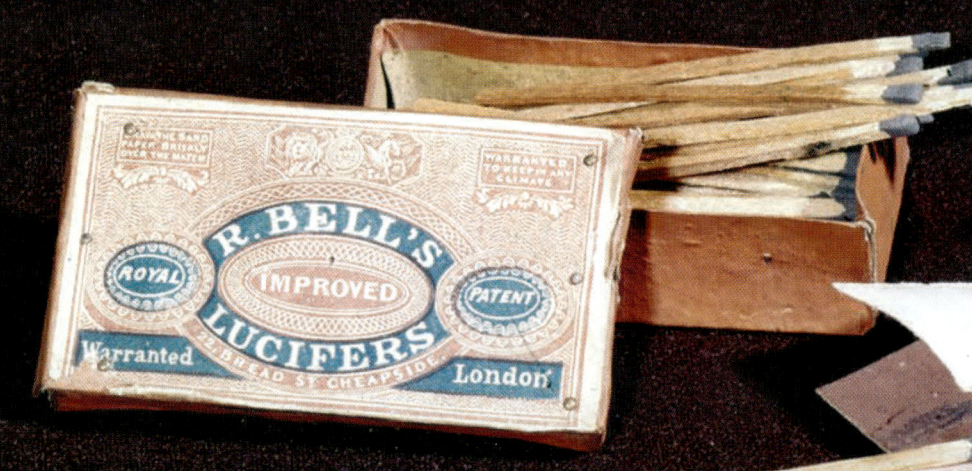

Walker called his creation **Friction** Lights. He mostly just sold it to his friends and family. However, the early matches could be dangerous. Sometimes the flame would drip, setting the user's clothing on fire.

THE DISCOVERY

Gustaf Erik Pasch created a better version of the match in the 1830s. His match used safer chemicals. But they still had their issues. They were very expensive to make. And the striking surface wore out quickly.

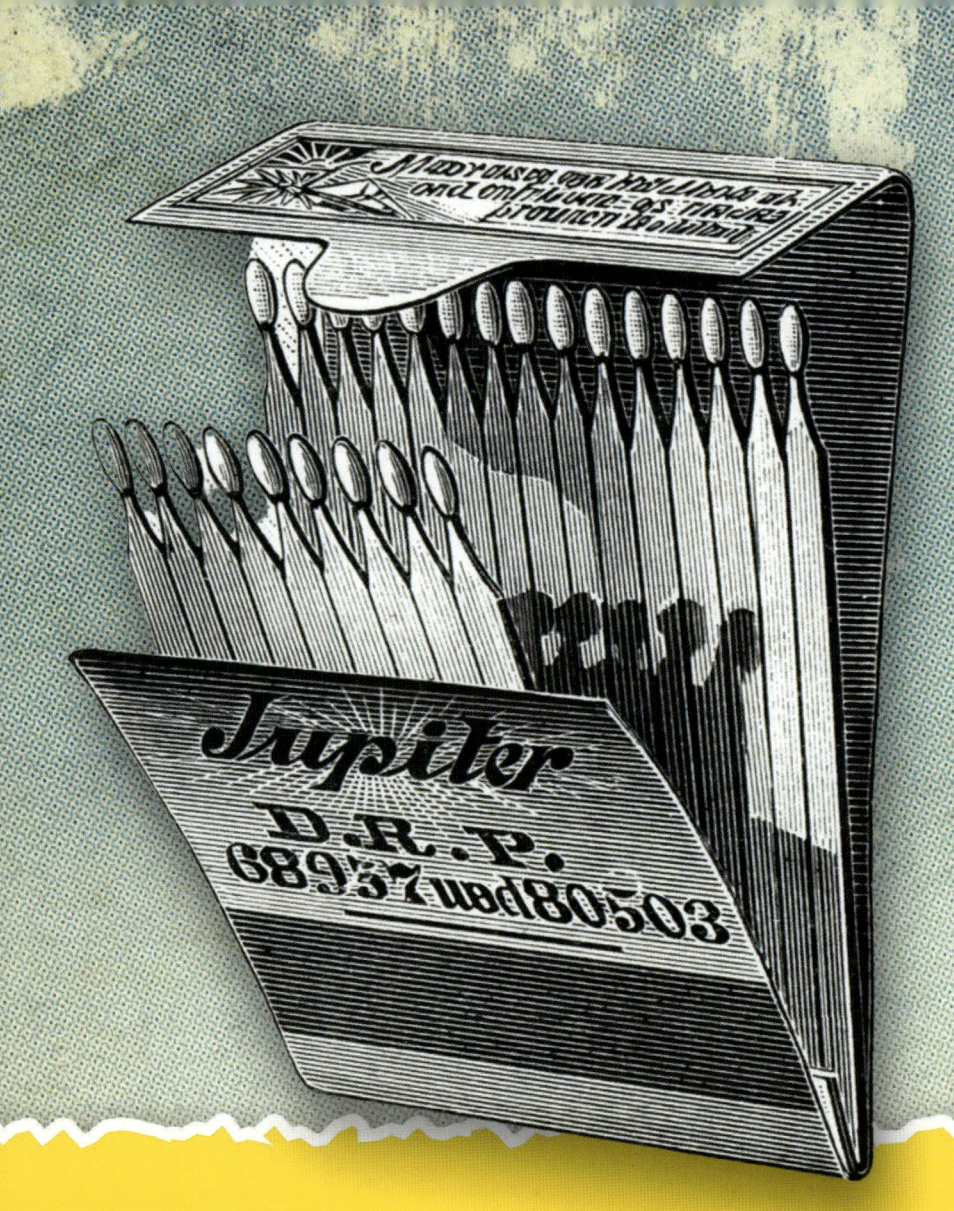

In 1845, J.E. Lundström began work to create a better type of safety match. His matches would only **ignite** on special, rough paper. However, many people still preferred strike-anywhere matches.

Strike-anywhere matches were cheap to make. The matches were used by soldiers during **World War I**. They quickly became popular around the world.

The match has been improved throughout the years. Different kinds were made for different uses. Common ones are safety matches, cigar matches, and fireplace matches.

Today, matches have been mostly replaced by lighters and other devices. However, matches remain an important tool for camping, hiking, and other outdoor uses.

THE FOOTPRINT

Matches continue to be a useful device in our daily lives. Their invention and **evolution** spark new ideas and **innovations**!

GLOSSARY

chemist – a person who conducts research and experiments with chemicals.

civilization – an advanced state of development of a society as judged by such things as having a system of government and laws, using a written language, and keeping written records.

evolution – the process of steady improvement.

friction – a force between two objects that are in contact with each other, sometimes creating heat.

ignite – to set on fire.

innovation – a new idea, method, or device.

mining – the process of obtaining useful materials, such as coal and iron, from the earth.

World War I – (1914–1918) a war fought in Europe. Great Britain, France, Russia, the United States, and their allies were on one side. Germany, Austria-Hungary, and their allies were on the other side.

ONLINE RESOURCES

To learn more about matches, please visit **abdobooklinks.com** or scan this QR code. These links are routinely monitored and updated to provide the most current information available.

INDEX

cigar matches 16

fireplace matches 16

Friction Lights 9

Lundström, Johan Edvard 12

Pasch, Gustaf Erik 10

safety matches 10, 12, 16

strike-anywhere matches 12, 15

uses 15, 16, 19, 21

Walker, John 7, 9

World War I 15